Contents

General Preface to the Series   iii

Preface   iii

1   Introduction   1
1.1 The birth and growth of tissue culture   1.2 Definitions of tissue culture terms

2   Tissue Culture Media   7
2.1 The essential features of a culture medium   2.2 Natural media   2.3 Synthetic media   2.4 Antibiotics   2.5 Sterilization of media

3   Containers and Techniques for Tissue Culture   16
3.1 The Maximow double coverslip method   3.2 Roller tubes 3.3 Flasks, bottles and Petri dishes   3.4 Chambers for microscopy   3.5 Organ cultures   3.6 Cleaning and sterilization of apparatus

4   Methods of Examining Cultures   26
4.1 Fixation and staining   4.2 Electron microscopy   4.3 Light microscopy   4.4 Time-lapse cinemicrography   4.5 Measurement of growth

5   The Basic Tissues *in vitro*   37
5.1 Epithelial tissue   5.2 Connective tissues   5.3 Muscle   5.4 Nervous tissue   5.5 Cell movement

Further Reading   58

General Preface to the Series

It is no longer possible for one textbook to cover the whole field of Biology and to remain sufficiently up to date. At the same time teachers and students at school, college or university need to keep abreast of recent trends and know where the most significant developments are taking place.

To meet the need for this progressive approach the Institute of Biology has for some years sponsored this series of booklets dealing with subjects specially selected by a panel of editors. The enthusiastic acceptance of the series by teachers and students at school, college and university shows the usefulness of the books in providing a clear and up-to-date coverage of topics, particularly in areas of research and changing views.

Among features of the series are the attention given to methods, the inclusion of a selected list of books for further reading and, wherever possible, suggestions for practical work.

Reader's comments will be welcomed by the author or the Education Officer of the Institute.

1977

The Institute of Biology
14 Queen's Gate,
London, SW7 5HU

Preface

The cultivation of animal tissue *in vitro* was first shown to be a significant experimental procedure in 1907. Since then, tissue culture techniques have been refined, extended and applied by an increasing number of biologists, and tissue culture is now one of the most fruitful methods in biological and medical research. This short account of the basic knowledge and skills required for the successful cultivation of animal cells and tissues is designed to help the reader to appreciate the scope and limitations of some of the most commonly used culture techniques. Perhaps their greatest virtue is that they allow us to study the structure and behaviour of *living* cells, and I have included a brief discussion of some of the ancillary methods, such as phase contrast microscopy, which must be used to obtain information of scientific value from tissue cultures. The observation and interpretation of the activities of cultured cells is, to me, a source of unending fascination, and I have tried to convey a little of this in the final chapter, which illustrates some of the characteristic features *in vitro* of cells from each of the basic histological tissues.

Leeds, 1977

J. A. S.

1   Introduction

1.1   The birth and growth of tissue culture

At the beginning of this century, biologists who were interested in the development of the nervous system held conflicting views about the way in which the axons of nerve cells were initially formed in the embryo. The dispute had been going on for a long time; in 1886 the embryologist Wilhelm His had postulated that the primitive embryonic neuron or neuroblast produced its axon by the outgrowth of a process from its cytoplasm, which continued to elongate until its advancing tip made contact with a peripheral sense organ or muscle fibre. This view was supported in 1890 by Santiago Ramon y Cajal, who studied the development of neurons by applying a silver impregnation technique to sections of embryonic nervous tissue. The opponents of His and Cajal adhered to the 'cell-chain' theory, which proposed that the axon was produced by the fusion of a continuous chain of initially separate cells extending from the neuroblast to the peripheral structure it was destined to innervate.

An American zoologist, Ross Granville Harrison, was deeply interested in the development of the peripheral nervous system. Having already written a number of scientific papers on the results of his research into the histogenesis of peripheral nerves in the embryos of fishes and amphibians, he had come to the conclusion that His and Cajal were probably correct but, in his own words, 'that the ordinary methods of histology were inadequate to answer definitely the question of the origin of the nerve fiber'. He decided that the best way to solve the problem was to devise a method 'by which the end of a growing nerve could be brought under direct observation while alive, in order that a correct conception might be had regarding what takes place as the fiber extends during embryonic development from the nerve center out to the periphery'.

With this end in mind, Harrison first dissected out fragments of the primitive spinal cord from frog embryos and put them into a saline solution, but the tissue failed to survive and disintegrated. Later, he tried a semi-solid medium, gelatine, but this was also unsuccessful. Then, in the spring of 1907, he devised a method which did succeed. Here is his first description of it, published in the same year:

'The method employed was to isolate pieces of embryonic tissue known to give rise to nerve fibers, . . . and to observe their further development. The pieces were taken from frog embryos about 3 mm long, at which stage, i.e., shortly after the closure of the medullary folds, there is no visible differentiation of the nerve elements. After carefully dissecting it out the piece of tissue is removed by a fine pipette to a cover slip upon which is a drop of lymph freshly drawn from one of the lymph sacs of an adult frog. The lymph clots very quickly, holding the tissue in a fixed position. The cover slip is then inverted over a hollow slide and the rim sealed with paraffine (Fig. 1–1). When reasonable aseptic precautions are taken, tissues will live under these conditions for a week and in some cases specimens have been kept alive for nearly four weeks. Such specimens may be readily observed from day to day under highly magnifying powers.'

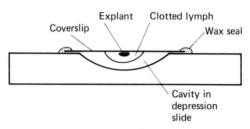

Fig. 1–1   Harrison's hanging drop culture method.

Not surprisingly, a problem Harrison had to overcome was the infection of his cultures by bacteria, which spoiled many of his first attempts. He persisted, and was eventually able to establish a reasonably aseptic procedure, although he found that 'the making ready of the apparatus consumes so much time, and the constant attention to the details of manipulation during operations is so fatiguing, that only a small number of preparations can be made in one day'.

Harrison's patience was rewarded; he succeeded in growing neurons outside the body in a medium (clotted lymph) where all possibility of contributions on the part of other living tissues was eliminated, and was able to see the outgrowth of axons from neuroblasts hour by hour, providing rigorous proof of the theory of His and Cajal (Fig. 1–2).

Harrison was not, in fact, the first person to keep cells alive outside the body. As early as 1885, Wilhelm Roux had explanted pieces of chick embryo into warm saline, where they survived for a few days; Arnold in 1885, and Jolly in 1903, had both transferred leucocytes (from frogs or salamanders) into saline or serum and had observed movement and division of the living cells. But Harrison is now generally regarded as the father of tissue culture because his experiments showed that, by adapting the hanging drop preparation on a depression slide—a technique already used by bacteriologists, it was not only possible to keep tissues alive *in vitro* for several weeks, but also that the procedure was a research method capable of making a fundamental contribution to biological knowledge.

Having succeeded, by ingenuity and patience, in developing a

The Institute of Biology's
Studies in Biology no. 82

An Introduction
to
Animal Tissue Culture

J. A. Sharp

M.D., Ph.D.

Senior Lecturer in Anatomy
University of Leeds

Edward Arnold

First published 1977
by Edward Arnold (Publishers) Limited
25 Hill Street, London W1X 8LL

Sharp, John Anthony
  An introduction to animal tissue culture.
  —(Institute of Biology. Studies in biology;
  no. 82).
  1.   Tissue culture
  I.   Title   II.   Series
  591'.072'4   QH585
  ISBN 0-7131-2644-2
  ISBN 0-7131-2645-0 Pbk

Printed in Great Britain by
The Camelot Press Ltd, Southampton

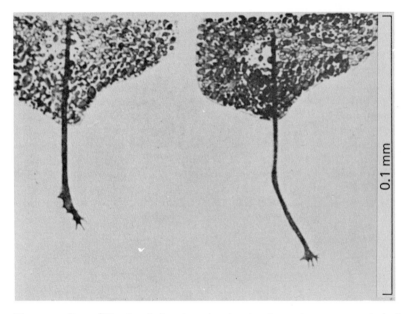

Fig. 1–2 One of Harrison's drawings showing the elongation, over a period of 25 minutes, of a nerve fibre *in vitro*. (Courtesy of the Wistar Institute.)

technique which enabled him to solve his immediate problem, Harrison himself played no further part in perfecting it, though indirectly through his published papers and lectures he made a valuable contribution by attracting the attention of other scientists to the potential importance of tissue culture. Indeed, the first major improvement in the method was introduced in 1910 by an American doctor, M. T. Burrows, who spent a few months of that year working in Harrison's laboratory in order to learn the technique. Burrows was interested in growing the tissues of warm-blooded animals. It was already clear that frog lymph left a good deal to be desired as a culture medium, partly because it did not produce a very firm clot and also because of the difficulty in obtaining it in sufficient quantities. Burrows' solution was to use chicken plasma to support and nourish explants of chick embryo tissues in hanging drop preparations. This proved to be much better than lymph, permitting good growth of nervous tissue, heart and skin.

Burrows then went on to collaborate with his colleague Alexis Carrel in an effort to extend tissue culture to mammalian tissues, and they soon succeeded in growing explants from adult dogs, cats, rats and guinea pigs, and also in growing malignant tissues. In addition, they showed that the life of cells *in vitro* could be prolonged by subculturing, that is by subdividing an established culture and transferring the living cells into fresh medium. Carrel and Burrows also discovered that, by mixing a proportion of embryo extract (tissue 'juice' extracted from minced chicken embryos) with the plasma, even better survival and growth of their cultures was achieved.

It was Alexis Carrel who was largely responsible for the next phase of refinement of tissue culture procedures and, having been trained as a surgeon, he concentrated on the application of rigid asepsis to the manipulation of cultures. Indeed, the methods which he evolved began to give other biologists the impression that tissue culture was an extremely laborious and expensive business. However, he did show that it was feasible to keep a strain of cells alive for 34 years by repeated subculturing, an achievement which owed much to his invention (1923) of the 'Carrel flask' (Fig. 1–3), which made it easier to avoid accidental infection of the cultures and reduced the number of manipulations involved in the maintenance of long-term cultures.

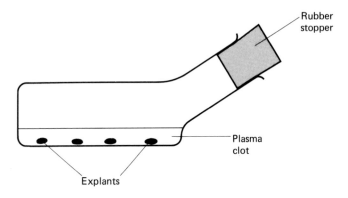

Fig. 1–3   The Carrel flask.

A different aspect of culture technique was investigated by two other Americans, W. H. and M. R. Lewis, who made the first attempt to replace natural media (plasma and embryo extract), whose composition could not be precisely defined, with synthetic media made up of known components. Their efforts, together with those of many others during the next 30 years, led eventually to the development of a number of synthetic media which can now be purchased in ready-made form. These have made it possible to reduce drastically the proportion of natural medium required and even, to a limited extent, to grow cells in a totally synthetic medium of precisely known composition.

The next major development in culture methods occurred not in America but in England, and made it possible to prepare cultures of a quite different kind in which the aim was to maintain small pieces of tissue, or entire embryonic organs, *in vitro* in such a way that their normal histological structure was maintained and the component cells were discouraged from growing out of the explant in a disorganized fashion. This approach, generally known as Organ Culture, was perfected by Dame Honor Fell at Strangeways Research Laboratory in Cambridge. The method, described in a paper in 1929, involved placing organ rudiments from chick embryos on the surface of a clot formed by mixing chicken plasma and embryo extract in a small watch glass enclosed in a Petri dish containing moist cotton wool (Fig. 1–4). The explants drew their nourishment from the underlying plasma clot and their oxygen from the air to which they were freely exposed, and the tendency for the

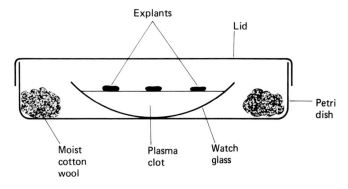

Fig. 1–4   The watch glass method for organ culture.

cells to migrate from the explants was minimized. By using this technique Fell and her collaborators were able to contribute a great deal to our knowledge about the development of bones and joints.

   In this brief introduction it has been possible to mention only a few of the many research workers who, during the last 70 years, have contributed to the improvement of tissue culture techniques and have brought them to bear on an increasing range of biological investigations. Some indication of the growth in the application of these methods can be gained from the fact that about 400 titles were listed in a bibliography on tissue culture published in 1927 (covering the two previous years), whereas the *Index of Tissue Culture* for 1974 contains over 30 000 titles of papers involving tissue culture which were included in the *Index Medicus* between January and December of that year. Examples of some of the most significant advances which have stemmed from the utilization of cultures of animal or human cells include the establishment, in 1956, of the true number of human chromosomes and the discovery shortly afterwards that Down's syndrome was due to the possession of an extra chromosome. In the field of virology the availability of cell cultures has simplified the cultivation of viruses and has made a vital contribution to the development of immunization against poliomyelitis, and more recently our understanding of the mechanisms which produce cell movement has gained immensely from the application of electron microscopy to cells grown *in vitro*.

1.2   Definitions of tissue culture terms

An inevitable result of this expansion was a tendency for confusion to arise about the precise meaning of the various terms used in tissue culture, and this led to the establishment of a committee on nomenclature, which produced a report on 'Proposed Usage of Animal Tissue Culture Terms' published in 1967. Some of the most commonly used terms are defined as follows.

Animal tissue culture: is concerned with the study of cells, tissues and organs explanted from animals and maintained or grown *in vitro* for more than 24 hours. Dependent upon whether cells, tissues or organs are to be maintained or grown, two methodological approaches have been developed.

Cell culture: this term is used to denote the growing of cells *in vitro* including the culture of single cells. In cell cultures the cells are no longer organized into tissues.

Organ culture: this term denotes the maintenance or growth of tissues, organ primordia, or the whole or parts of an organ *in vitro* in a way that may allow differentiation and preservation of the architecture and/or function.

Other terms in common use include the following.

Cell line: arises from a primary culture at the time of the first subculture, and consists of numerous lineages of the cells originally present in the primary culture.

Cell strain: derived by the selection or cloning of cells having specific properties or markers, which must persist during subsequent cultivation.

Clone: a population of cells derived from a single cell by mitosis.

Explant: an excised fragment of a tissue or an organ used to initiate an *in vitro* culture.

Monolayer: a single layer of cells growing on a surface.

Primary culture: started from cells, tissues or organs directly taken from organisms.

Subculture: the transplantation of cells from one culture vessel to another.

Suspension culture: a type of culture in which cells multiply while suspended in medium.

2 Tissue Culture Media

The individual cells in a living metazoan animal exist in an environment which is very precisely controlled. The surface membrane of the cell *in vivo* is exposed to extracellular tissue fluid, from which it obtains the substances essential to its survival such as water, inorganic ions, amino acids, vitamins, glucose, oxygen and so on, and into which it discharges the products of its metabolism. Certain attributes of the tissue fluid are maintained within narrow limits; thus in mammals its pH is held at 7.4, its osmotic pressure is equivalent to that of a 0.9% solution of NaCl, and it is normally sterile. Furthermore, the temperature of the cell's environment is regulated, in birds and mammals, with considerable precision.

The cell *in vivo* is very much dependent on other tissues and organs; on the liver, kidneys and lungs for the control of the composition, pH and osmotic pressure of the tissue fluid, on the haemopoietic tissues for defence against infection, and on the nervous system for the regulation of temperature. When living cells are excised from an animal they are deprived at a stroke of all these physiological supporting and protecting mechanisms. If they are to be kept alive *in vitro* they must then be provided with an artificial environment which resembles their milieu *in vivo* as closely as possible, in much the same way that an astronaut must be enclosed within a protective capsule and equipped with life support mechanisms if he is to survive in outer space.

2.1 The essential features of a culture medium

The composition and properties of the culture medium are the most crucial factors in achieving the successful cultivation of cells *in vitro*, and the medium must fulfil the following essential requirements.

1. Nutrients
It must provide all the salts, amino acids, lipids, carbohydrates, vitamins, etc., needed by the living cell.

2. Buffering capacity
It must contain non-toxic buffers to maintain the pH at 7.0–7.3 in spite of the production of acidic substances by the cells.

3. Isotonicity
The concentration of substances dissolved in the medium must render it isotonic with extracellular fluid. If the medium is hypertonic the cells will lose water and shrink, while if it is hypotonic they will take up water and swell.

4. Sterility

The medium must be devoid of micro-organisms which, if present, will find the conditions in the culture ideal for their rapid multiplication, and they will soon outgrow and destroy the living cells.

Media may be classified into two distinct categories: natural media, which are derived from the body fluids or tissues of animals, and synthetic media, which are precisely defined mixtures of substances dissolved in pure water.

2.2   Natural media

2.2.1   Cockerel plasma

It was seen in the previous chapter that the first successful tissue cultures relied upon natural media, and of these cockerel plasma soon became the most popular. It serves two purposes, providing the nutrients for either avian or mammalian cells and also, by forming a clot, it produces a network of fibrin which gives physical support to the explant and to the cells which grow out from it. Plasma from a cock is preferable to that from a hen, which is subject to considerable fluctuations in calcium content during egg production. Although mammalian plasma can also be used, it suffers from the disadvantage that it produces a clot which is less firm and more prone to detach from the coverslip. It has the additional drawback of forming a coarser fibrin network, impairing the optical clarity of the culture.

Cockerel plasma can be collected fairly readily by withdrawing blood from the wing vein of a healthy young adult bird, using a sterile hypodermic needle and syringe and taking care to sterilize the skin with alcohol before the needle is inserted. The main problem is to prevent the clotting of the blood, but this can be achieved by coating the interior of the syringe with a dilute solution of the anticoagulant heparin, and by transferring the blood quickly from the syringe to sterile tubes containing a little heparin and cooled in ice. The blood is then centrifuged, and the supernatant clear liquid plasma is pipetted into sterile glass containers which are sealed and stored in a refrigerator. When cultures are being set up, a drop of the liquid plasma is usually mixed with a drop of embryo extract, which causes the plasma to clot firmly within a few minutes.

2.2.2   Serum

Although plasma continues to be used in certain types of tissue culture, it has largely been superseded as a natural medium by serum. The species from which serum is obtained is not particularly critical, but humans, calves or horses provide convenient sources and, generally speaking, serum from young donors sustains better growth of cultured cells. Foetal calf serum is often preferred, and has the additional advantage for some experiments that it contains no antibodies (gamma globulins). Human foetal serum can be obtained by collecting blood from the umbilical cord and placenta after birth and is often used for maintaining delicate cultures of cells from the central nervous system. The preparation of serum is straightforward; blood is collected aseptically and allowed to clot, the clot is broken up and centrifuged, and the clear supernatant serum is transferred to sterile containers and stored in a freezer at − 20° C.

2.2.3   Embryo extract

Another natural medium which was widely used in the early days of tissue culture, but which has more restricted application today, is chick embryo extract. This provides a rich source of small molecular nutrients such as amino acids and nucleic acid derivatives and seems to stimulate the migration and mitotic division of cells *in vitro*. As already mentioned, it is often used in combination with cockerel plasma to produce a mixture having good supportive and nutritive properties. Chick embryos which have been incubated for ten days may be taken to prepare the extract. They are removed aseptically from the eggs, placed in a 20 ml syringe and disrupted by forcing them through the nozzle into a sterile tube. An equal volume of balanced salt solution is then added, the mixture is stirred, allowed to stand for an hour or two and then centrifuged. The supernatant liquid is collected and stored at − 20° C.

2.2.4   Collagen

There is one further substance which deserves to be mentioned in this brief outline of natural culture media, and that is collagen, which has become increasingly important in recent years for the cultivation of certain types of tissues and cells. It differs from the natural media so far described in that it has little or no nutritive function, its main purpose being to provide physical support for cells *in vitro*, and in this role it has a number of virtues. It was introduced as an alternative to clotted plasma which tends to be digested by living cells. Collagen is not destroyed in this way, and its use simplifies the maintenance of certain types of long-term culture. In addition it seems to provide a more congenial surface on which cells readily adhere and migrate, but its most interesting property is its ability to promote and maintain the differentiation of highly specialized cells, like those of the nervous system, muscle and the liver, significantly better than the other materials used in tissue cultures.

Collagen is, unfortunately, more difficult to prepare than plasma. The original method was based on a solution of collagen produced by soaking the tendons from rats' tails in dilute acetic acid. The solution was then dialysed in distilled water to remove the acid. A drop of collagen solution was placed on a coverslip and induced to gel by exposure to ammonia vapour. Finally, the coated coverslip was washed in a liquid culture medium before the explants were placed on its surface. However, it has proved possible to simplify this procedure, and collagen is now widely used for the cultivation of cells of the types mentioned above.

2.3   Synthetic media

2.3.1   Balanced salt solutions

Turning now to the synthetic media, we must first consider the relatively simple mixtures generally referred to as 'balanced salt solutions', which are not only useful in themselves but are also important because they form the basis of the more complex complete synthetic media. In searching for a saline solution which would be tolerated by living cells, the early workers in tissue culture drew on the experience of physiologists who, by trial and error, had already devised solutions for the perfusion of excised organs. Ringer in 1883 used a solution of the chlorides of sodium, potassium and calcium which would keep an excised frog's heart beating for several hours. In 1895 Locke changed the concentration of these electrolytes to suit the mammalian heart, and in 1910 Tyrode produced a more complicated mixture which contained sodium bicarbonate as a buffer, and glucose in addition to the usual electrolytes.

The compositions of two of the most widely used balanced salt solutions, devised respectively by Earle and by Hanks, are shown in Table 1.

Table 1   Balanced salt solutions

Ingredients	Earle (mg/l)	Hanks (mg/l)
NaCl	6800	8000
KCl	400	400
$CaCl_2.2H_2O$	264	185
$MgSO_4.7H_2O$	200	200
$NaH_2PO_4.H_2O$	140	—
Na_2HPO_4	—	47.5
KH_2PO_4	—	60
$NaHCO_3$	1680	350
Glucose	1000	1000
Phenol red	17	17

Apart from furnishing the inorganic ions essential to the life of all cells, these balanced salt solutions provide an aqueous environment in which the correct osmotic pressure is established, chiefly by the sodium chloride; the pH is maintained at 7.2–7.4 by the buffering action of the sodium bicarbonate, and a convenient source of energy is available in the form of glucose. The indicator, phenol red, is included at a concentration which is not toxic to the cells but is sufficient to give visual warning of a significant change in the pH.

When the two solutions in Table 1 are compared, the most noticeable difference is the greater quantity of sodium bicarbonate in Earle's saline. This is due to the fact that Hanks' saline is designed to be used in a sealed culture system containing air, while Earle's saline is intended for use in culture systems exposed to a mixture of 5% CO_2 in air. These differences stem from the fact that sodium bicarbonate is a relatively inefficient buffer; it tends to dissociate, releasing CO_2 into the atmosphere and hydroxyl ions into the medium, which consequently becomes too alkaline. Thus the buffering capacity of the culture medium is better maintained if the proportion of CO_2 in the gas phase inside the culture vessel is fairly high. The situation is further complicated by the fact that some types of cell produce considerably more metabolic CO_2 than others. Cells of the first type will tend to grow better if cultured in Hanks' saline exposed to air in a sealed container, where the CO_2 which they release into the enclosed air will help to slow down the dissociation of the sodium bicarbonate. Cells of the second type will generally do better in Earle's saline exposed to a gas mixture containing 5% CO_2, which compensates for their low production of CO_2.

These difficulties in the use of sodium bicarbonate have led to the introduction of different types of buffer, of which the most useful is probably a compound with the unfortunate name of N–2–Hydroxy-ethylpiperazine–N'–2–ethanesulphonic acid, more often referred to simply as HEPES. Briefly, the advantages of this compound are that it does not require an atmosphere enriched in CO_2; its pKa (i.e. the midpoint of its buffering range) is 7.3 at 37° C, hence it is a more efficient buffer than sodium bicarbonate at the temperature and near neutral pH required for tissue cultures. Its molecular weight is 238.3, and at a concentration of 20 mmol/l it is non-toxic to cells and produces a physiological osmotic pressure.

There is a variety of salt solution which is used specifically in the treatment of tissues with the enzyme trypsin or with the chelating agent EDTA in order to provide a suspension of separated, individual cells. Cells are more easily detached from each other in the absence of calcium and magnesium ions, and the actions of trypsin and EDTA are enhanced in solutions lacking these ions, such as the calcium- and magnesium-free version of Dulbecco's phosphate buffered saline.

2.3.2   Complete synthetic media

Although cells will remain alive for several hours in a balanced salt solution, such simple mixtures can do no more than permit them to survive for a limited time. Many more ingredients must be included in a synthetic or defined medium if it is to be capable of supporting not merely the survival but the proliferation of cells for significantly longer periods. Much time and effort have been expended by a considerable number of scientists in trying to achieve the ultimate goal of producing a completely defined medium in which cells will grow indefinitely. This ideal has not yet been realized, except in the case of a few highly selected and rather 'artificial' cell types, but a wide range of synthetic media are now available which, with the addition of only a small proportion of natural medium (usually serum), will supply all the requirements for the long-term growth of cells *in vitro*.

An example of a complete synthetic medium, one of several in general use at present, is Eagle's Minimal Essential Medium, and the specification for this is set out in Table 2. In common with other media of this kind, Eagle's M.E.M. is based on a balanced salt solution, usually either Earle's or Hanks', which is omitted from this table.

Table 2   Eagle's minimal essential medium.
(Balanced salt solution omitted.)

Ingredients	mg/l
L-arginine HCl	126.40
L-cystine	24.02
L-glutamine	292.30
L-histidine HCl.H$_2$O	41.90
L-isoleucine	52.50
L-leucine	52.50
L-lysine HCl	73.06
L-methionine	14.90
L-phenylalanine	33.02
L-threonine	47.64
L-tryptophan	10.20
L-tyrosine	36.22
L-valine	46.90
D-Ca pantothenate	1.00
Choline chloride	1.00
Folic acid	1.00
i-inositol	2.00
Nicotinamide	1.00
Pyridoxal.HCl	1.00
Riboflavin	0.10
Thiamin.HCl	1.00

The composition of this medium, published by Eagle in 1959, was determined by studying the nutritional requirements of pure populations of established mammalian cell lines. Eagle found that, if any single component was omitted, the cells degenerated and died, but the complete medium, together with the electrolytes and glucose supplied by a balanced salt solution, was sufficient for the indefinite propagation of cells *in vitro* in the presence of only 5–10% of serum. It will be seen that the medium consists of 13 amino acids and eight vitamins. The amino acids are the L–isomers (laevo-rotatory), since D–amino acids are not incorporated into mammalian proteins.

The precise contribution made by the small proportion of serum which is usually added to the synthetic medium is still not understood. It seems likely that certain protein molecules (probably α–globulins) present in the serum exert a beneficial effect on the cultures by promoting the attachment and spreading of the cells and by stimulating cell division, and it has recently been suggested that hormones in the serum also play a significant part.

It has already been mentioned that the maintenance of cells *in vitro* in a totally defined medium devoid of any natural components is a practical possibility at present for only a few specific cell strains; this has been achieved by establishing cultures initially in a mixture of synthetic medium and serum, then gradually reducing the proportion of serum over a period of time, thus encouraging the cells to adapt their metabolism until they are capable of surviving in a 100% synthetic medium. In this situation the environment of the cells is known exactly, which may be a prerequisite for certain types of experiment, although the composition of the medium will soon be altered by the cells, and the nature of such changes cannot be predicted accurately. For the great majority of experiments based on tissue cultures, the standard mixture of 90% synthetic medium with 10% serum offers the best chance of success.

Whatever the medium, it must of course be renewed at regular intervals; the frequency of such 'feeding' will vary, depending on the metabolism of the cells being cultured, the total volume of medium available to them, and its composition.

2.4   Antibiotics

Antibiotics such as penicillin and streptomycin may be introduced into tissue cultures at concentrations which are not toxic to living cells but are adequate to inhibit the growth of many of the bacteria which may accidentally contaminate the cultures. Potassium benzyl-penicillin may be safely used at a concentration of 100 units/ml, and is effective against Gram-positive organisms. Streptomycin sulphate will inhibit the growth of Gram-negative organisms, and is recommended to be used at a concentration of 100 μg/ml. There has been a trend recently towards the use of a single antibiotic (gentamicin) which is effective, at a concentration of 200 μg/ml, against both Gram-positive and Gram-negative bacteria. Gentamicin has two advantages over penicillin and streptomycin; it is more stable, and it inhibits the growth of certain bacteria known as mycoplasmas. These are particularly common contaminants of tissue cultures, and their presence is not so easily perceived as that of other bacteria.

There is a danger, however, that antibiotics may be used merely as an insurance against faulty aseptic technique. If aseptic procedures are rigidly observed, as they should be, antibiotics will not generally be needed unless the tissue to be explanted is unavoidably contaminated.

2.5   Sterilization of media

There are few media which will survive sterilization by heat; balanced salt solutions may be autoclaved, provided that sodium bicarbonate which, of course, loses CO_2 and is converted to sodium carbonate when heated, is reconstituted by bubbling CO_2 aseptically through the sterilized solution. The only practical method for sterilizing the majority of media, both synthetic and natural, is by filtration. Several types of filter are available for this purpose, but the most convenient is the membrane filter, which permits the solution to be passed through a thin membrane, with pores about 0.2 μm in diameter, which will trap any contaminating bacteria. A convenient (though not particularly cheap) variety of membrane filter is illustrated in Fig. 2–1.

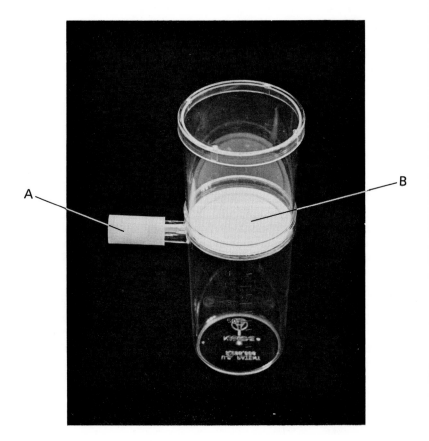

Fig. 2–1   A plastic membrane filter for the sterilization of liquid culture media. **A**, suction tube; **B**, filter disc. The filter disc has a pore size of 0.2 μm.

Balanced salt solutions, natural and synthetic media are now readily available from commercial sources, presterilized and ready for use. The cost of these is justified by the saving in time and effort which would otherwise be expended in preparing them, and by their uniformity and high quality.

3 Containers and Techniques for Tissue Culture

Cells will grow in almost any container, provided that it is made of non-toxic materials and can accommodate adequate supplies of nutrients and oxygen. In comparison with the unavoidable restrictions on the design of culture media, much greater freedom is possible in devising enclosures for tissue cultures, in terms of their size and shape and the nature of the materials used in their construction. Consequently very many different kinds of apparatus have been introduced, ranging from mere glass bottles to extremely complicated assemblies of pumps, tubing, chambers and reservoirs. The purpose of this chapter is to describe briefly some examples of the most widely used types of culture apparatus and to give an indication of their particular virtues and vices.

3.1 The Maximow double coverslip method

Two of the systems used in the early days of tissue culture, the hanging drop depression slide and the Carrel flask, were described in Chapter 1, and illustrated in Figs. 1–1 and 1–3. The first of these continues to have limited application, particularly for the cultivation of nervous tissue, but in a modified version known as the 'double coverslip' method introduced by Maximow in 1925. A major problem with the original hanging drop procedure of Harrison was the high risk of contamination of the coverslip carrying the culture. This is avoided in Maximow's method by establishing the culture on a small round coverslip, which is then attached to the centre of a large square coverslip by means of the surface tension exerted by a drop of sterile saline spread between the two. Both coverslips are then inverted over the cavity in a depression slide, and the edges of the large (upper) coverslip are sealed on to the slide with wax (Fig. 3–1). When

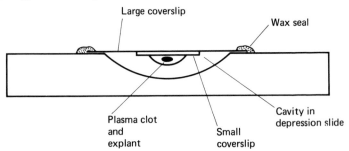

Fig. 3–1 The Maximow double coverslip culture method.

the culture is to be fed, the larger coverslip is detached from the slide, and
the smaller one carefully removed with sterile forceps; after the medium
has been replenished the small coverslip carrying the culture is then re-
attached to a fresh sterile large coverslip, which is sealed on to a sterile
depression slide. In this way, contamination of the small coverslip and
infection of the culture are avoided, and hanging drop preparations may,
with reasonable care, be maintained for long periods.

The components needed for the double coverslip method are simple
and cheap, and the cultures can fairly readily be examined with the low
powers of the microscope, but it is not suitable for detailed study of the
living cells at high magnifications. Furthermore, the volume of medium
in the hanging drop is small and it must be replenished frequently; it is
therefore a rather time-consuming business to maintain more than a few
of these cultures.

3.2   Roller tubes

The Carrel flask was a distinct improvement on the original hanging
drop method, and it led in turn to a still better device, the roller tube,
described by Gey in 1933. This is basically an ordinary test tube; several
explants are arranged in a row on its inner surface and covered with a
plasma clot. A few millilitres of liquid medium are then added and the
tube is sealed with a sterile, non-toxic rubber stopper (Fig. 3–2b). A
number of such tubes are then inserted into a slowly rotating drum within
an incubator maintained at 37° C. As each tube is carried round in the
drum the explants are alternately immersed in the liquid medium and
exposed to the air within the tube (Fig. 3–2c). Thus the cells are
automatically washed with nutrients and exposed to oxygen, and the
movement of the liquid medium ensures uniform distribution of its
components.

A refinement of this system has been given the rather misleading name
of 'the flying coverslip'. The procedure here is to attach the explants with
clotted plasma to a rectangular coverslip measuring 50 x 12 mm which
is inserted into a test tube containing liquid medium (Fig. 3–2d). This
makes it easy to remove the cultures from the tube for microscopical
examination or fixation and staining.

The roller tube is a convenient way of maintaining considerable
numbers of explants at a minimal cost in time and money; its chief
disadvantages are the very poor optical quality of the curved wall of the
tube, and the risk of infection when the rubber stopper is removed to
renew the medium.

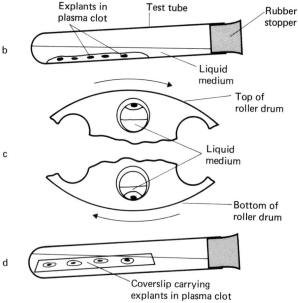

Fig. 3-2 The roller tube culture method. (**a**) General view of the apparatus. (**b**) Explants attached to the inner surface of the tube by plasma clot. (**c**) Exposure of the explants alternately to liquid medium and air as the tube is carried round in the drum. (**d**) Explants attached to a separate coverslip inserted into the tube (the flying coverslip method).

3.3   Flasks, bottles and Petri dishes

When large quantities of cells are required, as in some biochemical experiments and for the growth of viruses, large flasks or bottles are widely used. Here the optical quality of the wall of the flask is unimportant. The container is inoculated with a suspension, usually of cells of an already established strain. The number of individual cells per unit volume of medium must be established by counting the cells in a sample of the suspension, using a haemocytometer. The cells are then diluted with liquid medium to give a final concentration in the container of not less than 100 000 cells per ml. Equilibration of essential metabolites tends to occur between the cytoplasm of the cells and the surrounding medium, and if too few cells are present this can lead to serious loss of these substances from the cytoplasm with consequent impairment of the cells' growth.

The shape of the flask is not critical; conical (Erlenmeyer) flasks are convenient because of their stability. The cells will become attached to the surface of the glass as a monolayer, and will proliferate until they cover the available area. Their growth rate will then decline, but they will remain alive provided the medium is replaced every three or four days. Continued proliferation can be encouraged by detaching the monolayer of cells from the wall of a flask by gentle scraping or the use of trypsin, resuspending them in fresh medium and distributing them among an appropriate number of new flasks. Higher yields can be obtained from cells of certain strains by using 'shaker flasks'. The flasks are placed on a device which agitates their contents in a controlled fashion, thus keeping the cells permanently in suspension in the medium. Concentrations of 2×10^6 cells per ml can be achieved in this way, and a continuous supply of large quantities of living cells is provided.

For the growth of dissociated cells on a small scale, the Petri dish provides a convenient container. The ordinary glass type can be used, but the pre-sterilized plastic variety is particularly convenient and has the advantage that the bottom of the dish, on which the cells attach and grow, is flatter and more suitable for microscopic examination of the cultures. An inverted microscope, with the objectives below the stage and the light source and condenser above, is essential for this purpose. The thickness of the plastic is usually too great to permit high-power microscopy. Unless the lid of the Petri dish can be sealed water will evaporate from the medium when the culture is incubated at $37°$ C, and it may be necessary to place a number of dishes inside an air-tight container, or to use a humidified incubator.

3.4   Chambers for microscopy

The containers mentioned so far will allow examination of the cultured cells, if at all, only with the low powers of the light microscope. For those who use tissue cultures primarily to study in detail the structure and activities of cells *in vitro* the optical properties of the container are of paramount importance, especially since unstained living cells are virtually invisible unless a phase contrast or interference contrast microscope can be used. To make this possible, the cells must be grown in a chamber composed essentially of two coverslips held parallel to each other and separated by a distance of no more than 3–4 mm. Ideally, the coverslip to which the cells are attached should be 0.17 mm in thickness, because this is the optimum specified for the best high power objectives (other than the oil-immersion type). In the case of high power dry objectives with a numerical aperture (N.A.) in the range 0.65–0.95 (magnifications of 40× to 65×), the maximum deviation from this thickness should not exceed 0.02–0.005 mm respectively if full advantage is to be taken of the resolving power of the objective.

Provided these essential requirements are met, there is considerable freedom of choice in the size and shape of the chamber and in the materials used to support the coverslips and form the edge of the chamber. Cell biologists have demonstrated much ingenuity in contriving chambers to satisfy their individual needs, and dozens of different designs have been described in recent years. The tendency is for these to be referred to as 'perfusion chambers', but more often than not they are merely filled with medium which is renewed every few days, rather than having a continuous supply of medium perfused through them.

One example of a so-called perfusion chamber will be given here; it is a modification of a design first published by G. G. Rose in 1954. The component parts are illustrated in Fig. 3–3, and consist of two circular coverslips 50 mm in diameter, No. 1½ thickness (i.e. average thickness 0.17 mm), a rubber gasket, and two metal retaining plates with screws. The gasket has the same external diameter as the coverslips, and a central hole 25 mm in diameter; its thickness is 3 mm, but this may be reduced to 1.5 mm if necessary. It is made of silicone rubber, which is non-toxic, durable, can be heated to 180° C with impunity, is easily penetrated by hypodermic needles, and adheres firmly to the coverslips under the pressure exerted by the retaining plates. The metal plates are 50 mm square; each is 3 mm thick and has a central hole 25 mm in diameter. The edge of the hole is bevelled (on the side of the plate which faces outwards when the chamber is assembled) to permit the microscope objective the maximum possible freedom of movement across the coverslip. The plates may be machined from stainless steel, but aluminium alloy is satisfactory and cheaper.

The procedure for setting up primary explant cultures in this chamber is shown in Fig. 3–4. The explants are positioned near the centre of a coverslip; they may be attached with a drop of cockerel plasma, but for many tissues a strip of cellophane laid gently over the explants, its two

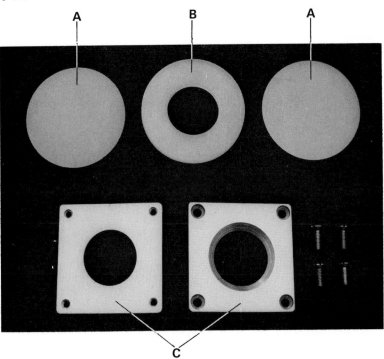

Fig. 3–3 Component parts of a simple perfusion chamber. **A**, coverslip; **B**, rubber gasket; **C**, metal plates.

ends projecting beyond the edge of the coverslip, provides a convenient way of holding the explants. The cellophane is cut from tubing of the type produced for kidney dialysis machines, sterilized in 70% alcohol, rinsed in several changes of balanced salt solution and finally in culture medium before being placed on the explants.

The rubber gasket is centred on top of the coverslip and cellophane, and is covered in turn by the second coverslip. The whole assembly is then inserted between the retaining plates, and the screws tightened until there is slight pressure on the two coverslips. Before the screws are fully tightened, the projecting ends of the cellophane are gently pulled to smooth out any wrinkles in the strip.

Narrow bore hypodermic needles are then pushed through the edge of the gasket on opposite sides until their tips just enter the central cavity, and liquid medium is injected through one needle while the chamber is tilted to allow the contained air to escape through the other. If the chamber is to be perfused with a continuous supply of medium, the two needles are left in position and connected to tubing; otherwise they are removed, and the medium is renewed when necessary by inserting fresh

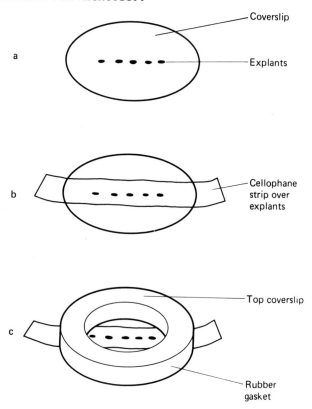

Fig. 3–4 Assembling the perfusion chamber using a strip of cellophane to hold the explants on the lower coverslip. (**a**) Explants placed on a coverslip. (**b**) Explants covered by cellophane strip. (**c**) Rubber gasket positioned over coverslip and cellophane and second coverslip placed on top of gasket.

needles, sucking out the old medium with a syringe and injecting a new supply.

When culturing a suspension of disaggregated cells the chamber is first completely assembled (without cellophane), needles are inserted and the suspension of cells in liquid medium is injected; Fig. 3–5 shows the chamber prepared for use in this way.

If the medium is renewed every two or three days, and the necessary aseptic precautions are observed, cells can be kept alive in this chamber for several weeks. It permits observation and photography with the highest powers of the light microscope. When the chamber has been incubated at 37° C for a few hours, the two coverslips adhere firmly to the silicone rubber gasket. The retaining plates may then be dispensed with, allowing the microscope objective to be focused on the contained cells

Fig. 3–5   Perfusion chamber assembled in metal plates with hypodermic needles inserted through the rubber gasket.

right to the edge of the central cavity. This adhesion can in fact be troublesome if it is wished to remove the coverslip on which the cells are growing, since it is almost impossible to detach it from the rubber without breaking it.

3.5   Organ cultures

The definition of an organ culture in the glossary (p. 6) stresses that its purpose is quite different from that of the tissue and cell culture techniques outlined above, which encourage the more or less disorganized growth of cells. In a successful organ culture the explants are provided with the minimum of nutrition necessary to keep them alive, but not sufficient to stimulate frequent cell division, and the conditions are such that cells do not migrate from the explanted fragment. Thus they retain *in vitro* the same position in relation to neighbouring cells that they had *in vivo*, so that the normal histological structure of the organ is preserved and the component cells may continue to fulfil their specific function.

Experience has shown that this can be achieved by supporting the explants at the interface between a layer of nutrient medium and an overlying volume of air or appropriate mixture of gases. The watch glass method employed by Dame Honor Fell has already been described (see p. 5); because of its cheapness and simplicity this has been widely and successfully used, especially for the culture of embryonic organs. Its chief disadvantage lies in the necessity for detaching the explants from the supporting plasma clot and transferring them to a new watch glass with fresh clot when feeding the cultures. A plasma clot is not always the most suitable type of support, since it may be digested and liquefied by the action of the cells in the explants, and it also rules out the use of a defined medium. Some of these difficulties can be overcome by incorporating a synthetic medium and serum in a 1 or 2% agar gel to support and nourish the tissue, but this does not avoid the need to disturb the explants when the medium is renewed. This problem has been overcome by supporting the explants with a metal grid of stainless steel mesh, folded to form a raised platform, on top of which the explants rest on a piece of lens paper, millipore filter membrane, or a thin layer of agar gel (Fig. 3–6). The mesh stands in a small Petri dish filled with liquid medium to the level of the top

Fig. 3–6 A plastic Petri dish containing a platform of stainless steel mesh covered with a strip of lens tissue for the support of organ cultures.

of the mesh so that the lens paper is just kept moist. Feeding is then simply a matter of sucking out the medium and replacing it, but care must be taken not to allow the fluid to cover the explants.

When Petri dishes are used as containers for organ cultures, they must be incubated in a humid environment to prevent evaporation of water from the medium, which was achieved in Fell's watch glass method by including some moist cotton wool in the Petri dish (see Fig. 1–4). This is not suitable for the metal bridge–liquid medium arrangement, and these cultures should ideally be placed in a humidified incubator. Failing this, they may be sealed into a chamber of the kind illustrated in Fig. 3–7, which will prevent evaporation of medium and can also be filled with a

Fig. 3–7   A chamber, designed by Trowell, for the incubation of organ cultures. The inlet and outlet tubes allow the interior to be filled with an appropriate gas mixture.

suitable gas mixture if necessary. Embryonic tissues survive quite well in air, but tissues from adult animals generally require pure oxygen (or, if the medium demands it, 5% CO_2 and 95% O_2). The explants must in any case not exceed 2 mm in diameter, otherwise the centre of each piece of tissue will be inadequately oxygenated and will become necrotic. Even under the best conditions, it is difficult to keep organ cultures alive and healthy for longer than seven to ten days (see Fig. 5–5, p. 41).

3.6   Cleaning and sterilization of apparatus

It is important that the materials which come into contact, directly or indirectly, with cultured cells should be 'chemically clean' as well as sterile. Immersion in nitric or hydrochloric acid is an effective method of cleaning glassware; detergents tend to persist on the surface of the glass, though 'Stergene' and 'Decon 75' may safely be used. Whatever the cleaning agent, it must be followed by thorough and repeated rinsing of the apparatus in distilled water.

For sterilization of glass, metal and silicone rubber, dry heat at 180° C for one hour is recommended, but autoclaving at 15 lb/in² (103.5 kPa) pressure for 20 minutes is an acceptable alternative.

4 Methods of Examining Cultures

Achieving success in the cultivation of animal cells and tissues *in vitro* is not an end in itself. To be of scientific value the cultures must be used in properly designed experiments, and further techniques must be applied to them to provide information about such things as the structure, activities, growth or biochemistry of the living cells. Some of these further techniques will be discussed in this chapter.

4.1 Fixation and staining

Cells which have been grown in such a way that they are attached to a coverslip (e.g. hanging drop, flying coverslip or perfusion chamber cultures) can be subjected to almost any of the standard histological stains by immersing the coverslip in a fixative solution, applying the chosen staining procedure, and then mounting the coverslip on a slide. This provides a permanent preparation suitable for routine light microscopy. Staining with the May-Grunwald and Giemsa combination is a useful method:

1. Rinse the culture in warm balanced salt solution.
2. Fix for 5 min in absolute methyl alcohol.
3. Stain 10 min in May-Grunwald.
4. Stain 20 min in Giemsa diluted 1 : 10 in distilled water.
5. Dehydrate rapidly in two changes of acetone.
6. Rinse in a mixture of equal parts of acetone and xylol.
7. Clear in xylol and mount on a slide.

Result: nuclei red-purple, cytoplasm and nucleoli blue.

In the case of organ cultures, the usual histological sequence of fixation, embedding, sectioning and staining is used for studying the internal structure of the explanted fragments of tissue (see Fig. 5–5).

Histochemical techniques for the detection of particular lipids, carbohydrates, nucleic acids or proteins (especially enzymes) designed for application to histological sections can be used on monolayers of cultured cells with little or no modification, and good results can be obtained because rapid and uniform fixation of the cells is possible. Cytochemical techniques for the detection of enzymes may be useful in studying changes in the metabolism of cells *in vitro*, and may also be helpful in distinguishing cells of a specific type which contain considerable amounts of a particular enzyme (Fig. 4–1).

Methods which have been developed for the staining of sections or smears with fluorescent dyes may similarly be adapted for use on tissue

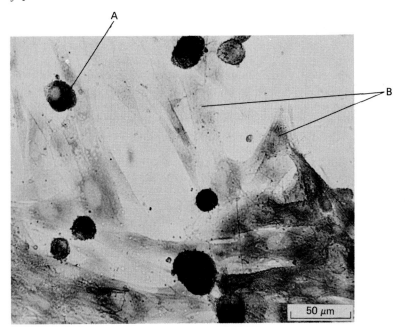

Fig. 4–1   A culture, containing macrophages (**A**) and fibroblasts (**B**), subjected to a cytochemical technique for the detection of an enzyme, non-specific esterase. The macrophages show an intense reaction for the enzyme, in contrast with the fibroblasts which contain relatively little.

cultures. These include the exposure of cultured cells to a solution of an antibody, labelled with a fluorescent dye such as fluorescein or rhodamine, for the detection of small quantities of a specific substance, and the demonstration of DNA and RNA by staining cells, fixed in alcohol, with acridine orange at pH 6.0. This stain gives rise to green fluorescence of DNA in the nucleus, and orange fluorescence of RNA in the nucleolus and the cytoplasm. The examination of such preparations ideally requires a microscope fitted with a mercury vapour lamp and the necessary filters to illuminate the cells with ultraviolet light, but useful results can be obtained with the cheaper tungsten halogen lamps in conjunction with an interference filter to cut out wavelengths above 500 nm.

4.2   Electron microscopy

The examination of ultrathin sections of cultured cells in the electron microscope has proved increasingly fruitful in recent years. Monolayers of cells growing on coverslips are frequently used for this purpose. The cells are fixed *in situ* with the usual sequence of glutaraldehyde and osmium tetroxide, dehydrated in alcohol, and embedded in Epon or one of the other plastic embedding media. There may be difficulty in separating the glass coverslip from the embedding medium when it has polymerized and hardened; this can be overcome by growing the cells on coverslips previously coated with a thin layer of carbon, or on coverslips composed of clear plastic. An alternative method is to use ordinary uncoated glass coverslips, embed the cells as usual, and then subject the polymerized embedding medium and the adherent coverslip to rapid cooling with solid CO_2, when the glass will readily separate from the plastic. A small area of the thin layer of Epon containing the cell or group of cells to be examined is then cut out, and sectioned on an ultramicrotome. Electron micrographs of excellent quality can be obtained in this way (see Fig. 5–14c) and, with care, it is possible to photograph or film an individual living cell *in vitro*, then to prepare serial ultrathin sections of the same cell and study them in the electron microscope. This combination of tissue culture with both light and electron microscopy has much to offer to the cell biologist who, by introducing biologically-active substances into the culture medium, can study their effects on the appearance and behaviour of living cells and correlate these with changes in the ultrastructure of the cells.

4.3 Light microscopy

The techniques which have been referred to so far are applicable only to cells which have been killed by the action of a fixative. While they are of undoubted value as sources of information about tissue cultures, they do of course defeat the primary objective of culture techniques, which is to make *living* animal cells available for study. But cells *in vitro* are transparent and virtually invisible under the ordinary light microscope, a fact which placed the early workers in tissue culture at a considerable disadvantage. The only method available to them for the detailed examination of living cells was dark field illumination, using a special substage condenser which illuminates the specimen obliquely so that the only light to enter the objective is that which has been scattered by the cell. Thus parts of the cell are highlighted against a black background. However, the optical properties of the container in which the cells are growing are extremely critical for this technique and, since only a very small fraction of the light contributes to the formation of an image, long exposures are required for photography unless the microscope is fitted with a photo-flash device.

4.3.1 Phase contrast

A solution of the problem of making living cells visible to the microscopist was provided in 1942 by a Dutch physicist, Frits Zernicke, whose studies on mirrors for astronomical telescopes led him to develop the theory of phase contrast. He realized that this could be applied to the microscope, and when instruments equipped for phase contrast became generally available after the war they were soon recognized as almost indispensable in tissue culture laboratories. Phase contrast microscopy continues to be of incalculable benefit in investigations on living cells, and its emergence from work in physics is a reminder of the continuing and increasing dependence of biology on the 'basic' sciences.

The additional components needed to enable a microscope to produce a phase contrast image consist of a black disc with a transparent annulus in the aperture of the substage condenser, and a 'phase plate' containing an annular groove in the back focal plane of the objective. An image of the clear annulus below the condenser is formed in the rear focal plane of the objective. This image must coincide with the annular groove of the phase plate. Briefly, light which is diffracted by optically dense structures in the specimen is retarded (i.e. rendered out of phase) in comparison with light passing through less dense areas. The phase contrast system permits rays of these two kinds to interact and thus converts phase differences (which are imperceptible to the eye) into amplitude differences which the eye can detect. Hence, with the usual type of equipment, denser structures in the cell appear darker in the final image. A more detailed (though simplified) account of the system is given in Fig. 4–2, and the dramatic increase in the amount of information contained in a phase contrast image, compared with what can be seen of a cell when the system is not used, is illustrated in Fig. 4–3.

The disadvantages of phase contrast microscopy are relatively minor. A high proportion (about 70%) of the light which enters the objective is absorbed by it, hence a fairly intense light source is needed and care must be taken to avoid damage to living cells which can be caused by illuminating them for long periods. Another concomitant of the method is the presence of a bright halo in the image immediately outside the edge of the cell, which can occasionally obscure details at the cell boundary.

There are two practical points about microscopes which are to be used mainly for the examination of cultured cells. It was mentioned in the previous chapter that, for cultures established in Petri dishes, an inverted microscope, with the objectives below the stage and the condenser above, was essential; an instrument of this type is also useful with perfusion chambers containing cells which do not attach to the glass coverslip (e.g. lymphocytes), because the chamber can then be examined with the cells still on the lower coverslip and there is no danger of them detaching and drifting out of focus. Several manufacturers produce inverted microscopes designed for tissue culture work, and one of these is shown in Fig. 4–4. The second point relates to the condenser, which should have a long focal length so that it can be used with culture chambers of reasonable thickness.

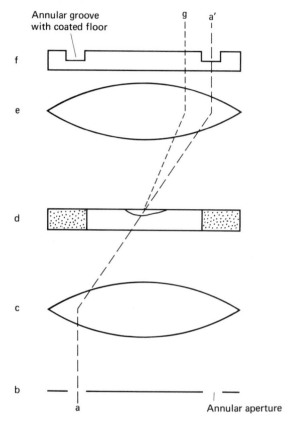

Fig. 4–2   The phase contrast microscope. A ray of light (a–a') passes through the transparent annulus (b) below the condenser (c), which directs the ray through a culture in the chamber (d). Here the ray emerges undeflected and enters the objective (e), where it is focused through the thin floor of the annular groove in the phase plate (f). The floor of this groove is coated to reduce the amplitude of the ray. A diffracted ray (g) also emerges from the culture; this initially has a lower amplitude than a–a' and is about a quarter of a wavelength out of phase with it. The diffracted ray (g) is then focused by the objective and traverses the full thickness of the phase plate so that it is further retarded, becoming about half a wavelength out of phase with a–a'. Above the phase plate the two rays are of approximately equal amplitude, and their interference produces an image in which dense structures in the culture appear darker than their surroundings.

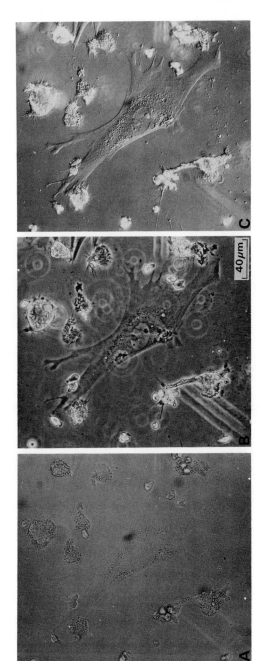

Fig. 4–3 A living fibroblast *in vitro* photographed (a) with ordinary optics, (b) with phase contrast, (c) with Nomarski interference contrast.

4.3.2 *Nomarski differential interference contrast*

This alternative method of increasing the visibility of living cells was introduced in 1955 by Georges Nomarski. Unlike earlier types of interference microscope in which two beams of light traverse the specimen, one passing through the cell and the other outside it, in the Nomarski system a single beam emerging from the objective is then split into two, separated by a very small distance, and these interfere with each other to produce an image in shades of grey. It is a more complex system than phase contrast, requiring a polarizer and a compensator prism to be fitted to the microscope condenser, with a beam splitter and an analyser inserted behind the objective, and no attempt will be made here to explain the underlying optical principles.

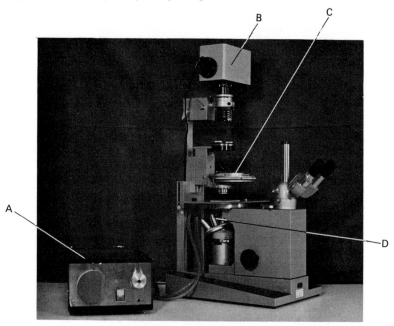

Fig. 4–4 An inverted microscope (the Reichert 'Biovert') for the examination of tissue cultures. **A**, thermo-circulator for warm stage; **B**, lamp; **C**, condenser; **D**, objectives.

When a living cell is viewed with Nomarski optics the image appears at first sight like an obliquely illuminated bas-relief of the surface of the cell (Fig. 4–3c), rather similar to the pictures of solid objects produced by a Stereoscan electron microscope. This appearance is deceptive, since in fact all the structures inside the cell which differ in optical density from their immediate surroundings contribute to the image, which is thus not merely a view of the cell's surface. The microscope can be adjusted so that cytoplasmic components which are more dense than their surroundings (e.g. granules, nucleoli, mitochondria) appear as 'elevations', while those which are less dense than the general cytoplasm (e.g. vacuoles) will then appear as 'depressions' in the image. Hence an interference contrast image is more difficult to interpret correctly than one in phase contrast, but it has certain advantages. The depth of focus is smaller with the Nomarski system so that different levels within the specimen can be studied by focusing through it; furthermore, very dense structures do not give rise to the halo effect which is a feature of phase contrast.

Nomarski interference optics have become generally available only within the last few years, and it is too early to assess their true value as an aid to the study of animal cells *in vitro*. It is unlikely that the Nomarski system will replace phase contrast as the best general purpose method of visualizing cells while they are alive, but it certainly provides a useful supplement to phase contrast, particularly if the microscope is equipped with both systems so that it is possible to turn quickly from one to the other.

4.4   Time-lapse cinemicrography

The movements of cells and of structures in their cytoplasm are often so slow as to be barely detectable even when long periods are spent looking at them through a phase contrast microscope. Time expended in this way, though full of interest, is unproductive scientifically and is inefficiently utilized, since an event of potential importance may easily be overlooked. It is essential to be able to record the activities of cells in a culture so that they may be studied repeatedly, subjected to analysis and measurement, and demonstrated to other people. For some purposes it may suffice to take a photomicrograph at intervals to show the important stages in a particular sequence of events, as in the illustration of mitotic division in Fig. 4–5. Such still photographs are useful for publication in a scientific journal, but they cannot provide a continuous record of the changes occurring in a culture over a period of hours. To achieve this a cine camera is needed, but if the camera is run at normal speed, not only will great quantities of film be consumed but many hours will be taken up in projecting and studying the film when it has been developed. Both money and time can be saved by applying the technique of time-lapse cinemicrography, which as the additional advantage of increasing the speed of movement of the cells when the film is projected, thus making their activities easier to apprehend. This is done, of course, by exposing a frame of the film in the camera (say) every five seconds and subsequently projecting the developed film at one of the standard speeds (16 or 24 frames per second).

A 16 mm cine camera is most often used because of the good quality of

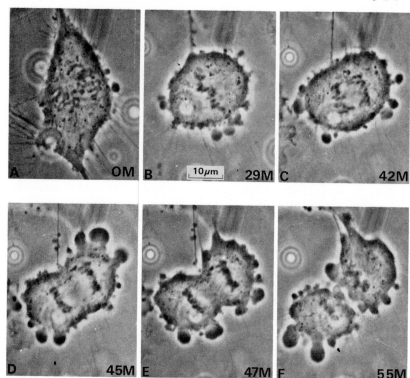

Fig. 4–5 A sequence of phase contrast photomicrographs illustrating the mitotic division of a living cell *in vitro*. The figures in the lower right-hand corner of each photograph indicate the time in minutes. **a** = late prophase; **b** = metaphase; **c, d** = anaphase; **e, f** = telophase.

the image which can be achieved with this frame size, but Super 8 mm equipment is now becoming available for time-lapse cinemicrography which is cheaper to operate and can produce an image of adequate quality for many purposes. The camera should preferably be driven electrically, and must be linked to a timing device which controls both the duration of each exposure and the length of the interval between exposures, permitting these to be varied according to the conditions. A rigid support must hold the camera so that no vibration is transmitted to the microscope. A beam-splitter interposed between the microscope eyepiece and the camera is essential to allow the field of view to be selected, the focus to be checked, and the intensity of the illumination to be monitored while the camera is operating. The final requirement is some means of maintaining the temperature of the culture on the microscope stage at 37° C. Equipment for time-lapse cinemicrography, based on an inverted microscope, is illustrated in Fig. 4–6.

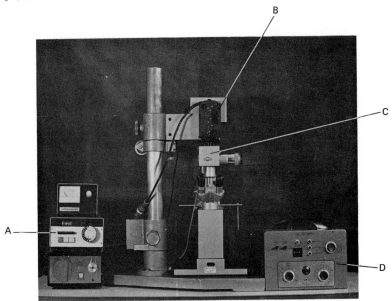

Fig. 4–6   Equipment for time-lapse cinemicrography. **A**, exposure meter; **B**, cine camera; **C**, beam splitter and eyepiece; **D**, camera control unit.

The quantity of information about the behaviour of cells *in vitro* which may be inherent in a time-lapse phase contrast cinemicrograph is very great; the same sequence of film can be viewed repeatedly, and each time some additional event may be noticed that had previously been missed. Hence these films must be studied with care. An analysing film projector, which permits individual frames at selected intervals to be projected, is helpful; for instance, it allows the outline of a cell to be drawn from every tenth frame, and knowing the exact time interval between frames, it is then possible to measure quite accurately the rate of movement of the cell and to plot its direction. The movement of a structure within the cytoplasm of a cell may similarly be plotted and measured. The combination of time-lapse cinemicrography and phase contrast optics thus forms a powerful method for the analysis of cell behaviour *in vitro*, particularly in the investigation of the effects of drugs or other biologically-active substances introduced into the culture medium.

4.5   Measurement of growth

There are several aspects of the growth of cells *in vitro* which may be taken as a basis for measurement. Thus growth may be assessed by counting the number of individual cells in a culture, or estimating the area occupied by the cells emerging from an explant, or determining the total quantity of protein or DNA in the cells, or their wet or dry weight. A choice must be made from these possibilities, depending on the object of the experiment and its design.

Reference has already been made in Chapter 3 to the haemocytometer as a cheap and simple procedure for counting the cells in a suspension culture. This can be done more quickly and accurately with an electronic cell counter, but such machines are expensive. Estimation of the total protein content or of the weight of a culture will provide no more than a rough indication of the number of cells present, since these quantities may both vary significantly without any change in actual cell number. A more reliable indirect method relies on the fact that the mean DNA content of individual cells does not alter, provided that polyploid cells are not being formed and that the distribution of cells between the stages of the cell cycle does not change. The total DNA content of a culture can be estimated by extracting the DNA and measuring it colorimetrically, using a spectrophotometer to compare the absorption of the extract with that of a sample of DNA of known concentration. Division of the total quantity of DNA by the amount present in a single nucleus then gives the number of cells.

Measurement of the area occupied by the outgrowth from an explant gives a result which reflects both the increase in number and size of individual cells and the extent of their migration from the explant. This can be done by projecting an image of the culture on to graph paper at a known magnification. The edge of the outgrowth is outlined on the paper, and the area can then be estimated by counting the number of squares within the outline. A less tedious procedure is to cut out the outline and weigh the piece of paper on an accurate and sensitive balance; by comparing this weight with the weight of a known area of the same paper, the area occupied by the outgrowth can be calculated. This method has the additional advantage that it does not interfere with the continued life of the culture, and the overall growth of the same culture can be measured repeatedly at regular intervals.

5 The Basic Tissues in vitro

The innumerable cells in the body of an animal can be classified into four distinct varieties. These are epithelial tissue, the connective tissues, muscle, and nervous tissue, and are referred to as the four basic tissues. An explant from any part of the body will contain cells belonging to one or more of these, but, as we have seen, when cells are cultured they are placed in a totally abnormal environment and the factors which previously determined their characteristic structure and activities *in vivo* no longer operate. Thus the loss of histological organization which occurs in cell cultures is often accompanied by loss of the specific structural and functional attributes which typified the individual cells in the intact animal, and cells which were recognizably different from each other *in vivo* may appear very similar in culture. E. N. Willmer has summarized this tendency as follows:

'Broadly speaking there are three main ways in which cells may behave *in vitro*, and according to their behaviour most tissue culture cells may be classed as (a) fibroblasts or mechanocytes, (b) epitheliocytes, or (c) amoebocytes or wandering cells.'

Willmer points out that there are certain cells (e.g. neurons and neuroglia) whose behaviour does not allow them to be included in this broad classification, but his general statement is true for many types of cell. In this final chapter the appearance and behaviour *in vitro* of some of the cells from each of the four basic tissues will be outlined, and the mechanisms which are believed to subserve cell movement will be considered briefly.

5.1 Epithelial tissue

Epithelial cells *in vivo* cover the external surface of the body (epidermis) and line the internal surfaces of the digestive, respiratory and genito-urinary systems, where they form layers of closely-fitting cells. They also exist as groups of specialized secretory cells in exocrine and endocrine glands.

5.1.1 Epitheliocytes in cell cultures

The inclination which epithelial cells have to come together as extensive sheets very often persists in cultures when a suitable surface is available for them, such as a coverslip, the floor of a Petri dish, or the interface between a plasma clot and liquid medium. A typical example of this pattern of behaviour is shown in Fig. 5–1. When primary explants rich in epithelial cells are cultured, such sheets will often emerge within a day or two at the periphery of the explant and gradually extend over the coverslip. The cells which constitute the sheet are in edge-to-edge contact from the outset and show little movement relative to each other. The sheet

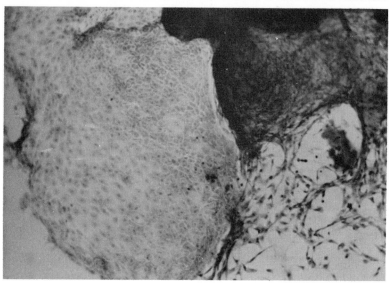

Fig. 5–1   A culture of kidney, stained by the May-Grunwald Giemsa method. A sheet of epithelial cells is visible on the left, with some fibroblasts on the right.

advances *en masse*, partly due to the traction exerted by the centrifugal movement of the cells at its free edge, partly by the emergence of additional cells from the explant, and also through the production of new cells within the sheet by mitotic division. The close fit between the constituent cells in an epithelial sheet *in vitro* is seen more clearly in Fig. 5–2. The individual epitheliocytes tend to have a smooth outline with relatively few blunt projections, and their nuclei contain one or two clearly defined nucleoli (Fig. 5–3). Their cytoplasm is moderately dense; filamentous mitochondria are distributed throughout, and there are occasional dark spherical bodies which may be lysosomes.

When ciliated epithelial cells are present in the explanted tissue (e.g. respiratory epithelium of the trachea or bronchial tree) they will often retain their cilia for several days after explantation and may give rise to sheets of cells covered with cilia which continue to beat rapidly. In cultures of glandular epithelium, clumps of secretory cells may be seen in the outgrowth (Fig. 5–4a), and their cytoplasm may contain phase-dark granules of stored secretion (Fig. 5–4b).

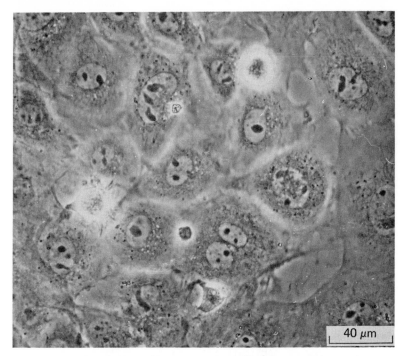

Fig. 5-2 A phase contrast photomicrograph of a sheet of epithelial cells. One cell is binucleate, and an adjacent one is in the early prophase stage of mitosis.

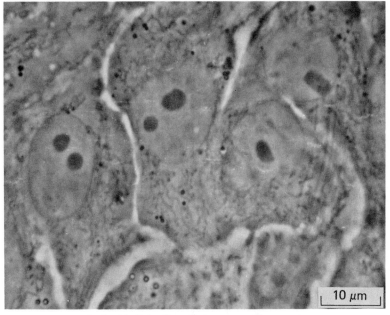

Fig. 5-3 Detail of epithelial cells at higher magnification. Phase contrast.

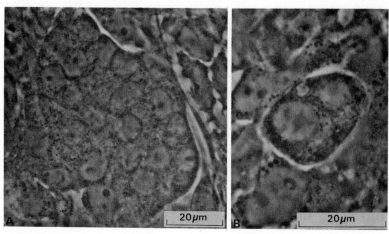

Fig. 5–4 (a) A group of secretory epithelial cells *in vitro*. (b) Detail of a similar epithelial cell at higher magnification, showing dark secretory granules. Phase contrast.

5.1.2   Epitheliocytes in organ cultures

Organs and tissues composed predominantly of epithelial cells have probably been studied in organ culture to a greater extent than any of the other basic tissues. Knowledge of the early differentiation of embryonic epithelial structures has been advanced by explanting them in organ culture. Thus the interactions between epithelial cells and their associated connective tissue during embryonic development have been clarified by establishing organ cultures of the epithelial components with or without their supporting connective tissue. Organ cultures of embryonic glands have been used to determine when they begin to release their specific secretion, and similar cultures of fragments of adult epithelial structures may be used to determine their response, in a controlled environment, to hormones, vitamins or other biologically-active substances. One series of experiments by Dame Honor Fell and her colleagues on organ cultures of skin was an investigation of the effect of introducing a large dose of vitamin A into the nutritive medium; this led to the conversion of the stratified squamous epithelium of the epidermis into a columnar ciliated mucus-secreting layer, which reverted to normal epidermis when the vitamin was withdrawn, a striking example of the ability of cells to change profoundly in response to an alteration in their environment.

The epithelial cells of the epidermis appear to survive better in organ culture than the connective tissue of the dermis (Fig. 5–5), perhaps because the epidermis *in vivo* lacks an intrinsic blood supply and is nourished by diffusion.

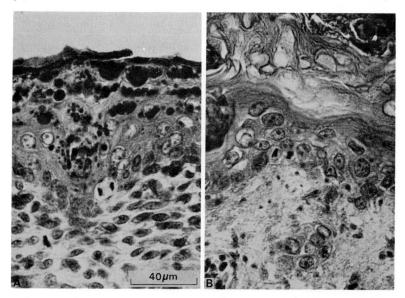

Fig. 5–5 Sections of foetal rat skin, (a) from a specimen taken immediately after the death of the animal, (b) from an explant after four days in organ culture. The epithelial cells in the epidermis of the explant are still healthy, whereas the connective tissue cells in the dermis have died. Haematoxylin and eosin.

5.2 Connective tissues

These include the 'true' connective tissues (fibrous tissue, cartilage and bone) characterized *in vivo* by large amounts of fibrous and amorphous extracellular materials, and the haemopoietic tissues (bone marrow and lymphoid tissue) which are densely populated by a variety of cells of the myeloid and lymphoid series, supported by a mesh of fine collagen and reticular fibres.

5.2.1 Fibroblasts or mechanocytes

The so-called 'fibroblast' (mechanocyte) is perhaps the commonest cell to be found in cultures from almost any source; it has been described as 'the weed in the tissue culture garden' because it tends to proliferate rapidly and choke the growth of other cells, especially when the medium is rich in nutrients.

Fibroblasts generally have a more angular shape than epitheliocytes and may be quadrilateral, triangular or fusiform, with several pointed processes (Fig. 5–6). They also differ from epithelial cells in that they emerge from the explant as an open network of cells rather than a tightly packed sheet (Fig. 5–1). When two living fibroblasts make contact they do not move across each other, but their movement either ceases or changes direction. This phenomenon, known as 'contact inhibition', which also occurs with epitheliocytes, not only tends to result in a monolayer of static cells but also inhibits their mitotic division. When malignant (cancer) cells are cultivated the control exerted by contact inhibition is inoperative, the cells pile up on each other and their rate of division is unchecked.

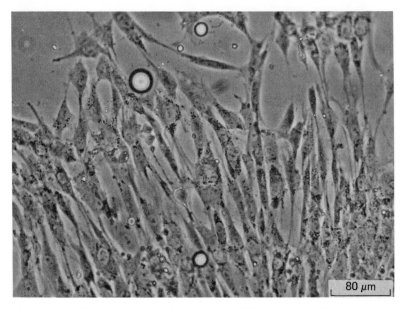

80 μm

Fig. 5–6   Typical fibroblasts ('mechanocytes') in the outgrowth from an explant of connective tissue. Phase contrast.

Although cells of this kind *in vitro* are usually referred to as fibroblasts, it must not be assumed that they are all derived from cells which, *in vivo*, were true fibroblasts engaged in the production and maintenance of collagen fibres. However, they will often produce both fibrous and amorphous (mucopolysaccharide) extracellular materials in culture.

5.2.2   *Macrophages*

Another connective tissue cell which is commonly encountered in tissue cultures is the macrophage (a typical amoebocyte or 'wandering cell') which originates *in vivo* from the haemopoietic tissues and enters the circulation as the immature monocyte. Monocytes emerge from the capillaries in all the tissues and organs of the body and become typical motile and phagocytic mature macrophages containing lysosomes for the digestion of the cell debris (or bacteria) which they ingest. Fig. 5–7 illustrates the appearance of an immature macrophage, which has a crescentic shape with active 'ruffled membranes' along its advancing convex edge. The cytoplasm contains clear droplets of liquid medium which have recently been taken up by the process of pinocytosis ('cell drinking'), first observed in cultured cells by W. H. Lewis and described by him in 1931 in the following words:

> 'The fluid globules which are taken up at the periphery of the wavy veil-like membranous pseudopodia seem to get caught or trapped in the folds of the ruffle, which probably then fuse around and completely enclose them. This process takes only a few seconds. The globules are at first irregular in outline but they soon become circular and probably spherical in form . . . and lie free in the cytoplasm of the pseudopod but surrounded, however, by a thin surface film which was originally a part of the surface membrane of the cell. One cannot see exactly how the folds enclose and fuse about a globule of fluid, but there can be no doubt about the fact that globules do actually get into the cell and then move rapidly towards the center.'

Since this first account, pinocytosis has been established as a regular feature not only of macrophages but of many other kinds of cell *in vitro*,

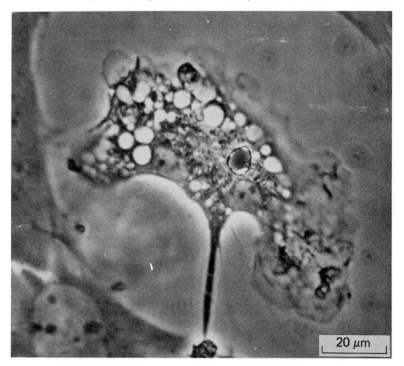

20 μm

Fig. 5–7   An immature macrophage *in vitro*. Phase contrast.

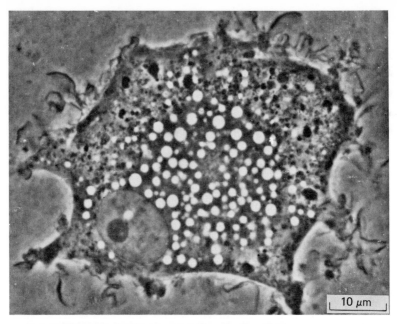

Fig. 5–8 A mature macrophage *in vitro*. Phase contrast.

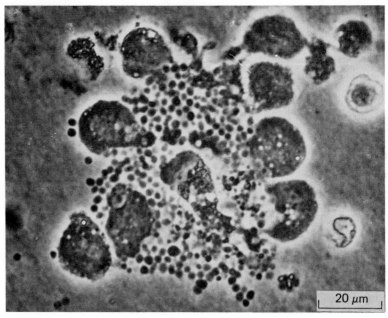

Fig. 5–9 A group of macrophages engaged in the phagocytosis of granules from a dead mast cell. Phase contrast.

particularly associated with the presence of ruffled membranes. Electron microscopy has revealed that it is not peculiar to cells in culture but is a common occurrence in many cells in the intact animal.

The appearance of a typical mature macrophage is illustrated in Fig. 5–8; there are many ruffled membranes at its margin, and small pinocytotic vesicles are visible in the peripheral cytoplasm on their way to join a group of larger vesicles in the centre. Lysosomes and phagocytosed debris are also visible. One of the most important functions of macrophages *in vivo* is the removal of the remnants of dead cells, a task which they can also be seen to perform *in vitro* (Fig. 5–9).

5.2.3   Haemopoietic cells

The general appearance of the cells in a culture of bone marrow is illustrated in Fig. 5–10; these constitute a diverse population, many of them being precursors of erythrocytes and granulocytes. Few of these cells are likely to be encountered in cultures of other tissues, with the exception

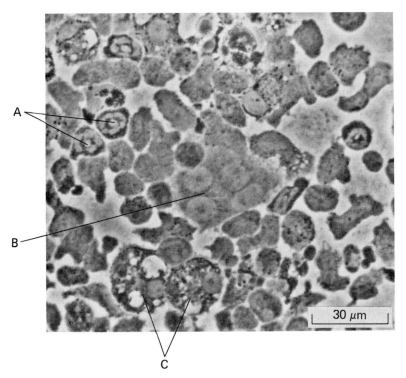

Fig. 5–10   A mixed population of cells in the outgrowth from an explant of bone marrow. Phase contrast. **A**, young neutrophil granulocytes; **B**, megakaryocyte; **C**, macrophages.

of monocytes and neutrophil granulocytes ('polymorphs'). The latter are easily distinguishable by their lobed nucleus and fine cytoplasmic granules; they are rapidly motile, constantly changing their shape and moving in a desultory and haphazard fashion.

The study of bone marrow *in vitro* has been advanced considerably by the introduction, in 1966, of a culture technique which promotes the proliferation and differentiation of the descendants of haemopoietic stem cells. It was discovered that these cells are stimulated to form colonies of macrophages or granulocytes by a substance (called 'colony-stimulating factor') released by peripheral blood leukocytes. The culture system used for this purpose consists of a Petri dish containing two layers of 0.5% agar gel in culture medium. The lower layer contains a 'feeder' culture of blood leukocytes, while the upper layer has bone marrow cells dispersed and suspended in the agar. Individual stem cells in the marrow culture proliferate in response to the stimulus provided by the 'feeder' layer, and each cell gives rise to a colony consisting of several hundred macrophages or granulocytes. This procedure is already proving useful in the investigation of patients with haematological diseases such as leukaemia.

5.2.4   *Lymphocytes*

The first cells to emerge from explants of lymph node, spleen or thymus (lymphoid tissues) are the lymphocytes. In time-lapse films these are seen to be capable of moving more rapidly than any other kind of cell. Like the macrophages and the neutrophil granulocytes, they are members of the amoebocyte group in Willmer's classification. Their movement tends to be intermittent; when travelling quickly they often assume a typical 'hand mirror' outline with small ruffled membranes in front of the nucleus and a tail of cytoplasm following behind (Fig. 5–11). After covering some distance in a purposive manner they may stop, become rounded and remain quiescent for a time before resuming their journey. Many observers have commented on the tendency which lymphocytes have to associate closely with other cells, sometimes actually entering them and moving around within their cytoplasm, a phenomenon which has been called 'emperipolesis' (inside round-about wandering). When lymphoid tissue which has recently been exposed to an antigen is explanted, lymphocytes may be seen moving persistently over the surface of macrophages, an activity, known as 'peripolesis', which could facilitate co-operation between lymphocytes and macrophages during the development of an immune response.

Many lymphocytes do not survive for more than a few days *in vitro*, but they are likely to be seen in small numbers in the initial outgrowth from almost any tissue. They are of fundamental importance in immunology, and much of our knowledge of their role in immune responses has been gained from experiments on cultures of lymphoid tissues.

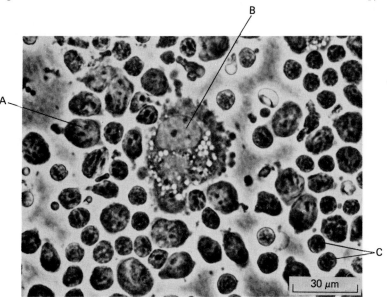

Fig. 5–11   Lymphocytes and a macrophage in the outgrowth from an explant of thymus. Phase contrast. **A**, lymphocyte with 'hand mirror' shape; **B**, macrophage; **C**, small lymphocytes.

5.2.5   Cultures of lymphocytes in cytogenetics

The culture of lymphocytes from the blood has become a routine procedure in genetics. When phytohaemagglutinin is added to such cultures, mitotic division of the lymphocytes is stimulated; subsequent addition of colchicine, which halts mitosis at the metaphase stage (see Fig. 5–20), causes the accumulation of large numbers of cells containing visible metaphase chromosomes. The cells are then exposed to a hypotonic solution in which they take up water and swell, and their chromosomes are dispersed more widely within their cytoplasm. Finally the cells are fixed, stained and spread on a slide under a coverslip. Photomicrographs can then be taken of the chromosomes in a number of the cells (Fig. 5–12). This is now a standard method of chromosome analysis, and is particularly useful in the investigation of genetic abnormalities in humans, since it not only allows the number of chromosomes to be counted accurately but also enables the individual chromosomes to be examined and classified according to their length and the position of their centromere.

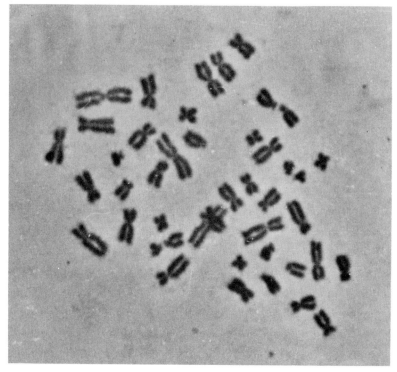

Fig. 5–12   A stained preparation of the chromosomes from a human peripheral blood lymphocyte.

5.3   Muscle

Tissue culture has been used to study all three varieties of muscle (smooth, striped and cardiac), but has been most fruitful in the case of striped (skeletal) muscle fibres. Comparatively little work has been done on smooth muscle *in vitro*; cardiac muscle from chick embryo heart was cultivated by M. T. Burrows as early as 1910 and he noted the persistence in culture of the inherent contractility of myocardial cells. The media used at that time caused the cells to lose their specific characteristics and tended to result in a rich growth of cells of the fibroblast type. These 'heart fibroblasts' became a popular subject for experiments on the living cell. More recently, trypsin dissociation has been used to prepare monolayers of individual cardiac muscle cells; when separated these beat irregularly and independently, but when they make contact with each other their contraction becomes synchronized and regular, illustrating the importance of intercellular communication in the action of the myocardium.

5.3.1   Skeletal muscle

When explants of chick embryo skeletal muscle are cultured, fully differentiated muscle fibres will appear in the outgrowth within a few days (Fig. 5–13) and will contract spasmodically for several hours before they degenerate, but the early stages in the development of the fibres are not clearly seen in such primary explant cultures. Again, the application of the trypsin dissociation technique has helped the detailed study of myogenesis *in vitro*. Monolayers of dispersed cells derived from embryonic skeletal muscle contain a mixture of fibroblasts and dense fusiform myoblasts (Fig. 5–14a), and after about four days in culture many of the myoblasts have fused together to form multinucleated myotubes (Fig. 5–14b). Electron microscopy of the cultured myotubes at this stage shows that myofilaments are being synthesized in the cytoplasm alongside the nuclei (Fig. 5–14c). Time-lapse films of these cultures reveal that the individual myoblasts produce myotubes by aligning to form chains of cells which then fuse at their tips. Thus, although fully differentiated striated muscle fibres do not survive for long *in vitro*, their embryonic precursors, the myoblasts, do retain their specific characteristics and activities in culture.

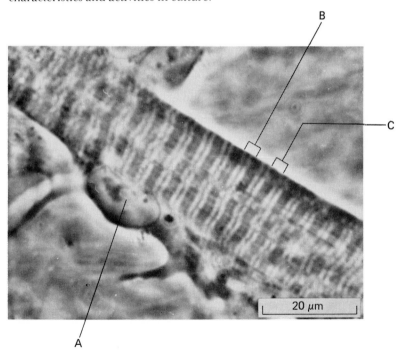

Fig. 5–13 A mature skeletal muscle fibre in a culture of chick embryo thigh muscle. Phase contrast. **A**, nucleus; **B**, 'I' band and 'Z' line; **C**, 'A' band.

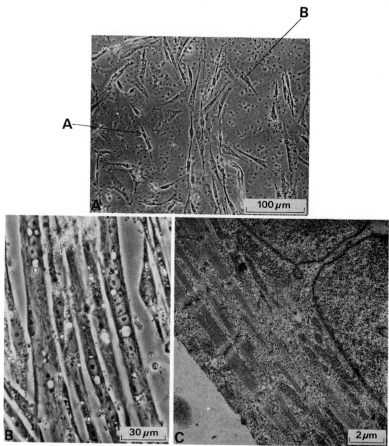

Fig. 5–14   The differentiation of chick embryo skeletal muscle *in vitro*. (a) Dissociated myoblasts (**A**) and fibroblasts (**B**) 35 h *in vitro*. Phase contrast. (b) Multinucleated myotubes produced by the fusion of myoblasts after 100 h *in vitro*. Phase contrast. (c) Electron micrograph of a myotube similar to those in (**b**). Myofilaments are visible alongside the nuclei. (Courtesy of Mr Jonathan Fear.)

5·4   Nervous tissue

Neurons and neuroglia (astrocytes and oligodendrocytes) have already been mentioned as examples of cells whose behaviour *in vitro* does not accord with Willmer's broad classification of mechanocytes, epitheliocytes and amoebocytes, since these highly specialized cells tend to retain many of their *in vivo* characteristics in tissue culture. Cultures of nervous tissue are particularly delicate; the composition of the nutrient medium and the nature of the apparatus in which they are grown are of greater importance than for most other tissues. Some of the older techniques, such as the roller tube and the Maximow double coverslip, continue to be the methods of choice for the maintenance of long-term cultures. The media which have been used tend to be complex mixtures composed of human placental serum, bovine serum ultrafiltrate, embryo extract, and a balanced salt solution with the glucose concentration increased to 600 mg/100 ml. Coverslips coated with collagen generally produce better cultures of nervous tissue than naked glass or plastic.

5.4.1   Central nervous system

When fragments of grey matter from the central nervous system are explanted, the first cells to emerge are usually fibroblasts and macrophages, the latter probably derived from the microglia. These are followed by astrocytes and oligodendrocytes, and later still neurons may appear in the outgrowth. It is often difficult or impossible to say with confidence whether a particular cell is a neuron or neuroglia, especially in cultures of nervous tissue from immature animals where many incompletely differentiated cells may be present (Fig. 5–15). Some

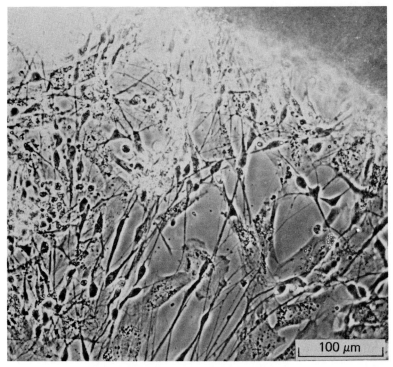

Fig. 5–15   A mixture of fibroblasts and dark, spindle-shaped immature neurons and neuroglia in the outgrowth from an explant of human foetal cerebral cortex. Phase contrast.

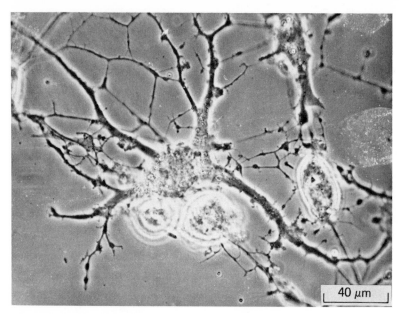

Fig. 5–16 A large neuron in a culture of human foetal cerebral cortex. It has dark granular cytoplasm and several branching dendrites. Phase contrast.

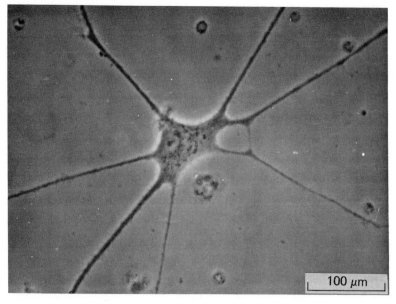

Fig. 5–17 An astrocyte in a culture of human adult cerebral cortex. Phase contrast.

neurons can, however, be identified because of their large size and their dense granular cytoplasm (Fig. 5–16). Astrocytes may also be recognizable by their long, straight processes and often eccentric nucleus (Fig. 5–17).

Much of the recent work on cultures of the central nervous system has been directed towards the study of the electrical activity of neurons and their ability to form synapses *in vitro*. By establishing cultures on collagen-coated coverslips by the Maximow method and then mounting a coverslip in a specially designed chamber it has been possible to record the electrical activity of living neurons *in vitro* and their responses to stimulation, using intracellular microelectrodes with a tip diameter of about 1.0 μm. This procedure has reached the stage where explants from different parts of the central nervous system are cultured on the same coverslip so that axons grow, for example, from an explant of spinal cord into an adjacent explant from the brain stem where they establish functioning synapses. Experiments of this kind require a great deal of care and technical expertise, but they may have much to offer in the analysis of the formation of specific synaptic contacts in the nervous system and the study of their responses to pharmacologically-active substances under the closely controlled conditions of tissue culture.

5.4.2   Peripheral nervous system

Cultures of dorsal root ganglia have the advantage, as compared with cultures derived from the central nervous system, that they contain a more limited range of different types of cell. Ganglia taken from chick or mammalian embryos at about the eighth day of development contain the easily recognizable cell bodies of sensory neurons, together with the capsule cells surrounding them, the Schwann cells associated with their axons, and some fibroblasts. Their comparative simplicity has made sensory ganglia a favourite object of study for those interested in the differentiation of neurons *in vitro*; they can be maintained by the double coverslip method for weeks or months, and provided there is good growth of capsule and Schwann cells the neurons will not only regenerate their axons but these will later become myelinated. Such myelinating cultures have confirmed the essential role of Schwann cells in the production of the myelin sheath and have even made it possible to see the slow rotation of the Schwann cell nucleus around the axon, a movement which had been predicted earlier from the spiral arrangement of the myelin lamellae revealed by electron micrographs of peripheral nerve fibres. The survival and growth *in vitro* of neurons from both sensory and sympathetic ganglia is promoted by a substance known as 'Nerve Growth Factor' (NGF) which can be extracted from the submandibular glands of adult mice, and when these ganglia are cultured in the presence of NGF the outgrowth of axons from the neurons is stimulated tremendously. Although NGF is useful to tissue culture workers as a means of improving the growth of ganglia *in vitro*, its mechanism of action and its role *in vivo* have still to be explained.

One of the most interesting features of a developing neuron is the 'growth cone' found at the tip of its advancing axon; this is difficult to identity *in vivo* but can readily be examined *in vitro* (Fig. 5–18). Electron

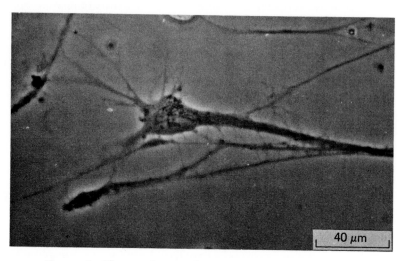

Fig. 5–18 The growth cone of an axon *in vitro*. Phase contrast.

microscopy of the growth cones of cultured neurons has provided a good deal of information about them which could not otherwise have been so readily obtained. Apart from mitochondria and membranes of the smooth endoplasmic reticulum, the growth cone contains a network of 5 nm wide microfilaments, particularly in the narrow projections ('microspikes') which extend out from the growth cone, while the axon itself contains, among other organelles, regularly arranged microtubules and 10 nm wide filaments (Fig. 5–19). To explain the potential significance of these microfilaments and microtubules in the growth of an axon, it is necessary to say something about the work which has been done on the analysis of cell movements in general.

5.5 Cell movement

Tissue culture has made an important contribution to the investigation of the mechanisms of cell movement by making it possible to observe in detail the behaviour of living cells using phase or interference microscopy, and to correlate this with their ultrastructure by preparing ultrathin sections of the cells for electron microscopy.

Following the elucidation of the role of actin and myosin filaments in

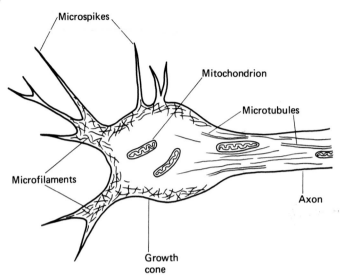

Fig. 5–19   The distribution of microfilaments and microtubules in the axon and its growth cone.

the contraction of skeletal muscle fibres, attention has been turned to the question of how non-muscle cells are able to maintain or alter their shape, to move about, and to produce the localized movements of the plasma membrane required in phagocytosis and pinocytosis. Improved methods of fixation for electron microscopy revealed the existence, in virtually all cells, of elongated cytoplasmic structures called micro-filaments and microtubules; the former are threads 5 or 10 nm wide, while the latter are more complex, appearing as hollow tubes 25 nm in diameter with a wall composed of 13 filaments. The microtubule filaments are polymers of a protein named tubulin. When non-muscle cells are exposed to specific antibodies, labelled with a fluorescent dye, against either actin or myosin, fluorescence is seen in areas of the cytoplasm known to be occupied by microfilaments, and there is other evidence which confirms the presence of actin in microfilaments. Thus actin and myosin are not confined to muscle fibres, but are probably present in all cells. The functions of microfilaments and microtubules have been clarified by culturing cells in the presence of either colchicine or cytochalasin B, which respectively cause the depolymerization of microtubules and microfilaments. The effects of these substances on the mitotic division of cells *in vitro* are summarized in Fig. 5–20; other experiments have demonstrated that the introduction of cytochalasin into a culture inhibits phagocytosis and the formation of ruffled membranes by living cells, and electron microscopy of these cells confirms that microfilaments normally present beneath the plasma membrane are disrupted by cytochalasin. The picture which is beginning to emerge as a result of investigations of this kind suggests that microtubules are implicated in the maintenance of the shape of a cell and in the transport of materials within its cytoplasm, while microfilaments are involved in movements of the plasma membrane, possibly through interaction between actin and myosin comparable with the sliding filament mechanism which operates in muscle fibres.

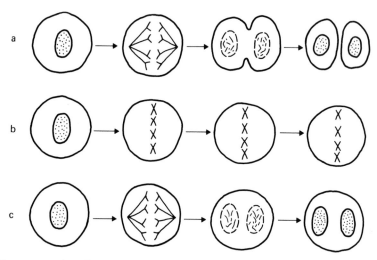

Fig. 5–20   The effects of colchicine and cytochalasin B on mitotic division; (a) normal mitosis; (b) colchicine prevents the formation of the microtubules in the spindle and the division proceeds no further than metaphase; (c) cytochalasin interferes with the process of cytokinesis (cytoplasmic division) involving microfilaments, resulting in a binucleate cell.

In the particular case of the axon and its growth cone, cytochalasin B *in vitro* causes the microspikes to disappear, the growth cone becomes rounded and the axon stops elongating. These changes are accompanied by disturbance of the organization of the network of microfilaments beneath the plasma membrane of the cone. Colchicine, on the other hand, does not affect the growth cone, but disrupts the longitudinal microtubules within the axon and causes it to collapse and retract into the cell body of the neuron. Thus the response of axons *in vitro* to the presence of cytochalasin or colchicine illustrates the way in which microfilaments and microtubules each make their different contribution to the ability of a cell to regulate its shape and initiate movements.

This book began with an account of Ross G. Harrison's efforts to grow embryonic axons *in vitro* so that he could watch their development away from the tissues which surrounded and obscured them *in vivo*. It has come full circle by ending with an outline of recent experiments in which tissue culture has made it possible to investigate the ultrastructural and molecular basis of axonal growth, using other techniques such as phase contrast and electron microscopy which were not available to Harrison. The technique which he launched in 1907 has become, and will continue to be, one of the most fruitful experimental methods in biology.

Further Reading

PAUL, J. (1975). *Cell and Tissue Culture*. 5th Edition. Churchill Livingstone, Edinburgh, An excellent practical laboratory manual of culture techniques.

WILLMER, E. N. Editor (1965). *Cells and Tissues in Culture* (3 vols.). Academic Press, London and New York. A comprehensive review of the results obtained by the application of culture methods to the various tissues and organs.

ROSE, G. G. (1970). *Atlas of Vertebrate Cells in Tissue Culture*. Academic Press, London and New York. A collection of phase contrast and Nomarski interference contrast photomicrographs, together with electronmicrographs, of a wide variety of cells *in vitro*.

KRUSE, P. F. and PATTERSON, M. K. Editors (1973). *Tissue Culture, Methods and Applications*. Academic Press, London and New York. A detailed and wide-ranging review of recent advances in tissue culture techniques, with many examples of their use in experiments on animal and plant cells.

BALLS, M. and MONNICKENDAM, M. Editors (1976). *Organ Culture in Biomedical Research*. British Society for Cell Biology Symposium I. Cambridge University Press. Intended to provide information and inspiration for those using or beginning to use organ culture methods.

Tissue culture bibliography

MURRAY, M. R. and KOPECH, G. (1953). *A Bibliography of the Research in Tissue Culture 1884 to 1950*. Academic Press, New York.

MURRAY, M. R. and KOPECH, G. (1965). *Current Tissue Culture Literature*. October House, New York.

Publications from 1966 onwards are listed in *Index of Tissue Culture*, now produced under the auspices of the Tissue Culture Association.